Attempt to Utilize Cryonics

*Reasons Why Utilizing Human
Cryopreservation Is Ultimately Desirable*

*Human Cryopreservation Is a Desirable
Potentially Life Extending Treatment*

Additional Books Available by Scott Everhart

Radical Life Extension: Psychological, Metaphysical, and Political Implications

Additional Books Available by Michael Ten

Outlaw Psychiatric Slavery (First Edition): Reasons for Outlawing Civil Commitment and the Insanity Defense

Radical Life Extension: Psychological, Metaphysical, and Political Implications

Attempt to Utilize Cryonics

*Why Utilizing Human Cryopreservation Is
Ultimately Desirable*

Scott Everhart
Michael Ten

Attempt to Utilize Cryonics: Reasons Why Utilizing Human Cryopreservation Is Ultimately Desirable

ISBN-13: 978-1530901999

ISBN-10: 1530901995

MichaelTen.com

ScottEverhart.com

DoNothingMedia.com

Acknowledgments and Dedications

I appreciate all those who have guided me in authentically beneficial ways and those who have motivated me to create more goodness and peace on Earth.

I appreciate all family, friends, and others that have helped me in many various ways.

I also want to specifically thank Sef Ramos for helping me to proof and polish this book.

I dedicate this book to all humans who attempt to live lives of authentic decency.

Table of Contents

Preface

I am writing this short book for all humans with the hopes that others may able to read it and be inspired to utilize cryonics so that the tragic loss of human life does not have to be emotionally processed or experienced.

Additionally, I am writing this book for myself and those that I know personally because I think that there is a possible chance that I may be able to live to be one-thousand years of age or older.

Why do I think this? I started believing this after I read the book *Ending Aging: The Rejuvenation Breakthroughs That Could Reverse Human Aging in Our Lifetime*. The book was written by Aubrey de Grey and Michael Rae.

Aubrey de Grey is a biomedical gerontologist. In the book, Aubrey de Grey notes that he thinks that there may be about a fifty percent chance that we may develop technologies to stop aging from killing humans in about twenty-five to thirty years. That book was published almost ten years ago. The book contains significant amounts of details as to how and why he thinks this might be possible. Aubrey de Grey advocates for a damage repair approach. There are seven fundamental types of age related damage that occur in the human body. In his book Ending Aging, Aubrey de Grey describes how we might be able to repair this damage so that the human body is rejuvenated, and the human life potentially extended. He has been covered in a wide variety of news publications and is renowned in various circles.

This book is not meant to be about anti-aging research specifically; it is meant to be about cryonics and cryopreservation technologies (which actually can perhaps be thought of as types anti-aging technologies). Quite frankly, I hope that individuals that I care about utilize cryopreservation technology, and that is partially why I am writing this book, in an attempt to energize and motivate the individuals who I personally like and care about on this Earth to utilize cryonics so that if I do happen live to be one-thousand years of age or older, I do not have to wait until I die to once again see the individuals who I know, like and care about.

I will also note that it seems that there is a small chance that the research that Aubrey de Grey is attempting to spear-head may be successful sooner than anyone suspects, and then all the individuals in my life may be able to utilize the radical life extension technologies.

Additionally, I think that I should note that I currently do not have arrangements for myself to be cryopreserved upon death, as it is legally defined, currently. However, in the future, as long as no one will force me to live if I do not want to, I hope to have arrangements made so that I will be cryopreserved upon death, as it is legally defined.

This is the second edition of this book. I have left the first version largely intact. I have added a section or so, and changed the writing style a bit. This second edition is meant to expand upon the first edition.

Thank you for taking the time to read this book. I hope that reading it is a useful and enjoyable experience.

Introduction

This book is about cryonics and cryopreservation technology. Specifically, this book is being written in an attempt to persuade you and those that you care about to, if possible, utilize cryopreservation technologies. Cryonics technologies are a metaphorical ambulance to the future.

The Alcor Life Extension Foundation is the main organization that I am aware of that deals with cryopreservation in The United States. Oregon Cryonics also offers cryonics services.

If financially possible, it seems that there are no significantly good reasons not to utilize cryopreservation after legal death.

This short book is meant to deal less with technical aspects of cryonics, cryopreservation, and regenerative medicine and more with the potential philosophical, ethical, and spiritual aspects of the aforementioned topics.

There are two main types of cryopreservation. There is neuro-preservation where just the brain and/or head is vitrified and preserved, and then also whole-body preservation, where the whole body is cryopreserved and vitrified.

It should also be noted that certainly, no one should be forced to utilize cryopreservation after death. Utilizing cryonics technology should be completely voluntary. Additionally, perhaps there will be a time when governments will offer cryopreservation to citizens for free upon death. Of course, future research certainly must be conducted in order to find ways to potentially revive individuals who have been declared legally dead and that have undergone cryopreservation.

Perhaps in fifty to eighty years' time technologies will be developed to revive and heal those that have been cryopreserved. Hopefully this is and will happen as soon as possible.

Of the following sections of this book, one considers reasons why cryopreservation may be undesirable. This is done to try and unmask for you why utilizing cryonics is ultimately the most desirable potential course of action. Of course, you are encouraged and free to decide on this philosophical and potentially moral issue yourself. To utilize or not to utilize cryopreservation technologies is ultimately a personal decision. Hopefully all individuals will embrace the possibility of life, and will attempt to utilize cryonics technologies.

Non-coercive persuasion should be used to convince those you care about to utilize cryonics.

I want this book to be something that can be given as a gift in order to help persuade those you care about to utilize cryonics.

I want to empower you and those that you care about to make the most informed decisions about cryonics as possible.

Radical life extension is another idea related to cryonics, which should be explained here. Radical life extension is the idea that we may be able to extend the human life span so that humans might live indefinitely. Radical life extension is not the main topic of this book, at least not directly.

Radical life extension relates to cryonics in the sense that if radical life extension technologies are developed in more than twenty years from the present, cryonics may be a way of saving one's life until radical life extension technologies are developed.

Why Cryopreservation Is Desirable

There are multiple reasons why utilizing cryopreservation technologies is desirable.

The following is one reason why utilizing cryonics may be more than just slightly desirable. There are a set of theories called Strategies for Engineered Negligible Senescence (SENS) that was laid forth in the book *Ending Aging: The Rejuvenation Breakthroughs That Could Reverse Human Aging in Our Lifetime* by Aubrey de Grey and Michael Rae. There are a variety of mainstream scientists who find these theories to be plausible. SENS Research Foundation seems to be the main organization that is attempting fund and conduct research related to SENS. If SENS research is successful, then it seems that it may be possible that individuals who are revived from cryopreserved states may be able to utilize biomedical treatments that enable them to have healthy and biologically young bodies that are free from age related immobility, pain and so forth.

Utilizing cryonics may be desirable because overpopulation may never have to be a problem. Harold White a researcher with NASA is attempting to conduct research that relates to warp field mechanics and may potentially lead to a potentially functioning warp drive somewhere down the road.[1] Additionally, many exoplanets are being discovered every year, and it is simply a matter of time before habitable planets are discovered. It does seem possible that significantly practical methods of colonizing other planets are never discovered. However, many things once thought impossible, or at least never thought possible have been shown to be possible. Over a hundred years ago, some individuals never thought human flight would ever be possible and that attempting to make it happen was foolish. They were wrong.

Another reason why cryopreservation may be desirable is that it seems theoretically possible to create a life worth living at any age. At any age it is possible for life to feel worthwhile and purposeful. Of course, as things currently stand, if one is a super-centenarian (someone over the age of 110), doing activities which can help to create a meaningful life may simply be physically difficult to do. For example, it is likely difficult to go snowboarding (or do other potentially enjoyable activities) at the age of 115. The longest confirmed living human who has ever lived is Jeanne Louise Calment. She died in the year 1997 at the age of 122. Relatively speaking, out of the roughly seven billion humans on Earth, there are quite a few who have lived more than 110 years of age. Logically and statistically, it seems that some of these individuals were likely quite miserable at the end of life. However, it seems likely that some of the super-centenarians that are also quite happy and content to be living. Of course, no adult should be forced to live, and much less forced to live indefinitely, if the technology to live indefinitely ever becomes available. The point is, aside from some objectively diagnosable neurological disorder, there is no proof that any human has the neuropsychological inability to create and experience a life worth living. This is a reason why cryopreservation should be desirable, it will likely still be possible to create a life worth living in the future, if revived from a cryogenic vitrification.

Cryonics can be thought of as similar to other already widely utilized medical interventions, and therefore may then logically be just as desirable as those. It is not unheard of to hear of stories where someone is declared clinically dead and then revived. Someone might indicate that they were dead for a certain amount of minutes. If cryonics is ever successful, then perhaps it can be thought of as being technically dead for a certain number of years, rather than minutes. If we find it morally and ethically desirable to try and revive someone who is clinically dead after a couple minutes, if possible, then it also seems equally ethically and morally desirable to try and bring someone back from clinical death after any number of years. This seems to be another reason to support the utilization of cryonics technologies for life extension purposes. Of course, this is not an apples to apples comparison. It can perhaps be metaphorically, thought of as an apricot to apriplum (a plum and apricot hybrid fruit) comparison, rather than an apples to oranges, or apples to apples comparison.

Finally, cryonics may be like a metaphorical ambulance to the future.

Cryonics may enable one to utilize future technologies that can revive someone who we would currently consider legally dead.

If radical life extension technologies are successfully developed, then cryonics may be a way to rejoin future generations of family. What could be better than potentially being able to spend time with multiple generations of family?

We all may eventually meet up in Heaven, however, why not attempt to rejoin as a family on Earth?

From a religious perspective, cryonics may be desirable. The Lord's Prayer states, "On Earth as it is in Heaven." Cryonics may enable the revival those who are now thought of as being legally dead. No one knows for certain how technologies will progress in 80 to 100 years. Cryonics may help to create or establish Heaven on Earth in some small way by decreasing the net amount of death that happens.

Even from a purely secular humanistic perspective cryonics may be desirable. Cryonics may effectively decrease the amount of death that happens, presuming that those cryopreserved are able to be revived in fifty to eighty years or more.

Cryonics can be viewed as a potentially lifesaving medical treatment. What we view as legal death now may not be viewed as legal death in the future.

Cryonics can only be legally done on those who are legally, technically, and officially dead. Humans can only undergo cryopreservation after legal death, and not before. All cryonics organizations are required to adhere to this standard.

We do not know what technology will be like in 100 or 200 years. Technology may advance to the point of being able to revive those who are considered legally dead and then cryopreserved. This will hopefully be the case, and of course, the sooner the better.

Why Cryopreservation May Be Undesirable

It seems important to examine potential objections to why utilizing cryopreservation technologies may not be desirable and the counterpoints to the potential objections for utilizing cryonics. Issues such as cryonics should be examined from both sides so that a fully informed decision about the topic can be made.

There may be multiple reasons why someone does not want to be cryopreserved upon death, as it is legally defined. Of course, no adult should be forcibly or coercively cryopreserved (nor forced to live for that matter). It will be quite good if more individuals choose to be cryopreserved and vitrified after death, as it is legally defined. This is so because cryopreservation is a metaphorical ambulance to the future. Cryopreservation can be thought of as a medical procedure. We currently define legal death in a way that could change in the future. The greater number of people cryopreserved is a greater number of lives potentially and saved in the future.

It is possible to try and utilize theological or spiritual justifications as to why cryopreservation should not be done upon what is officially considered death. However, these will likely not withstand any serious and well thought-out logical counterpoints. For the most part, all of the most popular interpretations of all the major religions and theologies on Earth support and promote life and living. Cryonics also has the potential to support life and living.

Perhaps one is just curious about what happens after life. That is a fair curiosity. It may be possible to see cryopreservation as an undesirable way to potentially delay authentic death.

Perhaps one has had a hard and mostly unpleasant life as they perceive and experience it. Then one might hope that death is better than life and not want to continue living. This is a fair reason to reject the use of cryonics. However, the counterpoint to this is that the experience of living can always be improved and it is never too late to enjoy life.

Additionally, it seems that it might only be desirable to be revived from a cryopreserved state if one could have a healthy physical body. It seems logical to not want to be revived if one will be in a body that is decrepit and produces the experience of pain, discomfort, and/or immobility.

Individuals can have stipulations attached to any potential for revival from a cryonics state.

It seems reasonable, if one passes away at the age of ninety-five to not want to be revived if upon revival from a cryopreserved state, one once again has the body in the condition of a ninety-five-year-old human. However, science may eventually be able to reverse human aging processes using principles and applications of biomedical gerontology and regenerative medicine.

From a theological perspective, for Christians, one might reason that cryopreservation stops or delays one from going to Heaven. Heaven is a great concept. How nice a perfect place it may be. This may be a reason to want to reject cryopreservation. If one has children, a potential justification for utilizing cryopreservation may be to be able to see one's own descendants before they die. Additionally, Christianity is a faith that is often devoted to helping others. Being cryopreserved and potentially revived at a later time can enable an individual who is Christian to potentially come back to Earth and continue to do good work.

It seems possible that one may not want to be cryopreserved due to concerns that overpopulation of Earth may occur. This is a potentially valid concern. However, it seems that potential solutions to significant problematic overpopulation can likely be found. If humans are able to utilize creativity, research, ingenuity, effort and resources to figure out how to fly, to put humans on our Moon, to create nuclear reactors and nuclear weapons and so on and so forth, then it seems that with those same methods of innovation and problem solving, humans can likely figure out a solution to potentially problematic overpopulation.

Some individuals who are religious reject all modern medical treatments. It seems that at least adults should have the option of doing so. This may be one reason why one may choose to not utilize cryopreservation. Persuading such an individual to utilize cryopreservation technologies may require some sort of mild to significant religious conversion. Such conversion is likely not worth the effort it might require. Although, in countries where individuals have religious freedom, individuals who support cryonics want to certainly use time and energy to try and convert these individuals to a theological position where they desire to be cryopreserved, or at least do not mind the idea of cryopreservation.

More Considerations

If cryopreservation upon death is utilized on a massive scale, problems may theoretically arise. For example, if millions of billions of humans are cryopreserved before technologies can be created to revive those and hopefully also regenerate the bodies of those who have been cryopreserved, then it seems possible that the physical space required on Earth to house (so to speak) all of the cryopreserved biological tissue of humans might become a problem. A potentially solution to this might be to store cryopreserved specimens in space or on other planets. This potential problem is a long way off.

It seems possible that science may never progress to the point where individuals that are cryopreserved can ever be resuscitated. This however seems unlikely if human creativity and ingenuity is relentlessly applied to finding a solution. Perhaps figuring how to revive individuals who are cryopreserved may take hundreds, thousands, or millions of years or more. However, it seems that eventually, at some point, a solution will be found with regards to how to do it.

It seems that suicide should not be effectively illegal. This may give individuals greater incentive to utilize human cryopreservation, in the sense that, if suicide is effectively legal, then cryopreservation may be less likely to potentially seem like a trap (that traps humans into living on Earth or in this Universe). Suicide should never be encouraged though.

Additionally, it seems that we should attempt to peacefully, non-forcefully, and non-coercively persuade individuals to not engage in suicide and help give individuals good reasons to live. This may involve more than effectively helping individuals solve problems in life that otherwise would be potential reasons for individuals to commit suicide. Creating a life worth living is likely possible for all humans.

Economic and Political Considerations

Eventually, cryonics services should be available to all individuals who want them.

This means publicly funding cryonics services.

The practicality of this may be decades off. However, making cryonics technologies available to all who want them is ultimately the moral thing to do, providing that it can be done in an economically feasible and peaceful way.

Cryonics should be a choice. No coercion or force should be involved in relation to cryonics services, ever.

Persuading Those You Care About

If you are attempting to persuade those you care about to utilize cryonics, there are multiple considerations to take into account.

Do not be coercive. Coercion has too many undesirable unintended consequences to be useful.

Use emotional and intellectual reason and persuasion. Sometimes emotional persuasion is more powerful than intellectual reasoning.

Sometimes it might not be possible to persuade others and acceptance would be the best course of action.

Slowly persuading others may be useful. It might be beneficial to try and slowly chip away at resistance to utilizing cryonics.

Remember, utilizing or not utilizing cryonics is a personal decision and we can only do so much to help others make a fully informed decision about it.

Radical life extension technologies may be developed. Cryonics technologies may be a good backup strategy incase radical life extension technologies are not developed in a timely fashion.

Cryonics may be a way of enabling one to spend time with future generations of family on Earth.

Cryonics may be the technology that will enable your loved ones to see you again.

Cryonics may be a way to meet back up with future generations of family.

Conclusion

This book exists in an effort to persuade individuals to utilize cryopreservation technologies in greater numbers.

I want to empower you so you can make more informed decisions about cryonics, and how to communicate it to others.

This book should be seen as a starting point. It seems likely that a rather long book might be written about why cryopreservation upon death as it is legally defined is desirable.

If one is cryopreserved, it makes sense that one should be able to have the cryopreservation expire after a certain amount of time. Perhaps one might want to stipulate that if researchers are not able to find out how to revive and regenerate their body after fifty, one-hundred, a thousand or five thousand years, then at that time, they will be buried or cremated.

Perhaps one may want to stipulate that if technology becomes available to revive cryopreserved individuals, that they only want to be revived if their body could also be regenerated so that they are the biological age of a human in their twenties (for example).

I hope that this book has helped to persuade you

to want to utilize cryonics technology if possible. If you are interested in utilizing cryonics, I recommend that you contact organizations (like Alcor Life Extension Foundation) that can help make it happen for yourself at some point in the future if needed.

It is possible to be cryopreserved after legal death. Please contact any of the aforementioned cryonics organizations if you are interested in having arrangements made to be cryopreserved when necessary. It is important to make arrangements before one's technical and legal death happens so that one can undergo cryonics procedures as soon as possible after legal death occurs and bodily systems cease functioning.

If you like this short book and feel that it is valuable, please recommend it to others.

Resources

Organizations

Alcor Life Extension Foundation – Alcor.org
Oregon Cryonics – OregonCryo.com
SENS Research Foundation – SENS.org
Methuselah Foundation – MFoundation.org
Buck Institute – BuckInstitute.org

Books

Ending Aging: The Rejuvenation Breakthroughs That Could Reverse Human Aging in Our Lifetime

Fatal Freedom: The Ethics and Politics of Suicide

Suicide Prohibition: The Shame of Medicine

Why God Won't Go Away: Brain Science and the Biology of Belief

More

Warp Field Mechanics 101. ntrs.nasa.gov/archive/nasa/casi.ntrs.nasa.gov/20 110015936_2011016932.pdf. This is a paper written by Harold White. It is available to read and access for free online.

Notes

1. White, Harold, Paul March, Nehemiah Williams, and William O'Neill. "Eagleworks Laboratories: Advanced Propulsion Physics Research." Johnson Space Center, 5 Dec. 2011. Web. 17 Aug. 2013. ntrs.nasa.gov/archive/nasa/casi.ntrs.nasa.gov/20110023492_2011024705.pdf.

Afterword

Thank you for reading this book. I hope that it has motivated you to think differently about death from human aging.

Please leave an honest review of this book where you are able to.

Additionally, please sign up for my mailing list so that you can stay up to date on when I release future books. You can sign up for the mailing list at MichaelTen.com/Subscribe or at DoNothingMedia.com/Suscribe or just visit my website at MichaelTen.com.

You can also subscribe to Scott Everhart's list at ScottEverhart.com/Subscribe or just visit his website at ScottEverhart.com.

www.ingramcontent.com/pod-product-compliance
Lightning Source LLC
Chambersburg PA
CBHW020715180526
45163CB00008B/3096